Contents

Introduction   1

The Link Between Emissions and Climate   2

Potential Impacts of Climate Change   5
- Effects on the Physical Environment   7
- Consequences for Biological Systems   11
- Impacts on the Economy and Human Health   12

Policy Implications of Uncertainty   14

Endnotes   18

Figures

1. Historical and Projected Climate Change   5
2. Illustration of Changes in Averages and Extremes in Temperature and Precipitation   6
3. Historical Climate Patterns in the Continental United States, 1971 to 2000   8
4. Historical and Projected Climate Change, With and Without an Illustrative Policy Applied   15
5. Uncertainty in the Climate's Response to Rising Atmospheric Concentrations of Carbon Dioxide   16

Box

1. The Current State of the Art in Climate Modeling   3

Potential Impacts of Climate Change in the United States

Introduction

Human activities around the world—primarily fossil fuel use, forestry, and agriculture—are producing growing quantities of emissions of greenhouse gases, other gases, and particulates and are also greatly altering the Earth's vegetative cover. A strong consensus has developed in the expert community that if allowed to continue unabated, the accumulation of those substances in the atmosphere and oceans, coupled with widespread changes in patterns of land use, will have extensive, highly uncertain, but potentially serious and costly impacts on regional climate and ocean conditions throughout the world.

This paper summarizes the current state of scientific understanding of the potential effects of projected changes in climate and related developments. The paper describes the wide range of potential impacts, including changes in: seasonal weather patterns; the amount and type of precipitation; storms and sea level; regular climate fluctuations; ocean acidity; ecosystems and biodiversity; agriculture, forestry, and fishing; water supply and other infrastructure; and human health. The discussion focuses mainly on projections of impacts in the United States but also refers to impacts elsewhere that could be particularly severe and could indirectly affect the United States. The paper draws from various primary and secondary sources, particularly the *Fourth Assessment Report of the Intergovernmental Panel on Climate Change* (IPCC), a major national assessment recently released by the National Science and Technology Council in the Executive Office of the President, and a number of surveys published by the U.S. Climate Change Science Program, which integrates executive-branch research on climate and global change.[1]

The paper emphasizes the extensive uncertainty about the future growth of greenhouse-gas emissions and resulting climate-related developments, and the implications of that uncertainty for climate policy. Uncertainty arises from several sources, including limitations in current data, imperfect understanding of physical processes, and the inherent unpredictability of economic activity, technological innovation, and many aspects of the interacting components (land, air, water and ice, and life) that make up the Earth's climate system.

Uncertainty does not imply that nothing is known about future developments, but rather that projections of future changes in climate and of the resulting impacts should be considered in terms of ranges or probability distributions.[2] For example, some recent research suggests that the median increase in average global temperature during the 21st century will be in the vicinity of 9° Fahrenheit (F) if no actions are taken to reduce the growth of greenhouse-gas emissions. However, warming could be much less or much greater than that median level, depending on the growth of emissions and the response of the climate system to those emissions.

The uncertainties surrounding the extent of climate change and its effects on human welfare complicate policy choices because any given emissions scenario could result in much more or much less dramatic changes in climate (and consequent impacts) than are expected. A full evaluation of any policy requires an assessment not only of its most likely effect on climate impacts but also of how it might shift the probabilities of impacts that are unlikely but would be very severe if they occurred. Because of that uncertainty, policymakers might wish to consider taking more action as a hedge against those severe outcomes than they would just to address the expected or most likely outcome.

The Link Between Emissions and Climate

Over the past century, researchers have developed an increasingly sophisticated understanding of the climate system through direct observations of the system, statistical analyses of those observations, and, more recently, simulations of the system using computer models. (See Box 1 for a discussion of climate modeling.) According to current understanding, as energy from the sun is absorbed by the Earth's climate system and radiated back into space, greenhouse gases increase the amount of energy temporarily held in the lower atmosphere, keeping the Earth's surface warmer than it would otherwise be.[3] Those gases include carbon dioxide (CO_2), methane (CH_4), nitrous oxide (N_2O), and several man-made gases containing fluorine and chlorine. Soot in the atmosphere also absorbs incoming energy and tends to warm the lower atmosphere, whereas aerosols tend to exert offsetting cooling effects.[4]

Since the onset of the Industrial Revolution about two centuries ago, the growth of emissions from human activities has resulted in rising atmospheric concentrations of greenhouse gases, other gases, and aerosols.[5] The accumulation of greenhouse gases and other warming substances has been the dominant influence contributing to climate change, triggering an irregular but accelerating warming of the Earth's surface and various consequent changes.[6] By current estimates, the average global temperature (currently about 58.0°F, or 14.4° Celsius) has most probably risen by about 1.4°F since the mid-19th century.[7] However, that estimate involves substantial uncertainty. There is roughly a 5 percent chance that the warming has been less than 1.0°F and an equal chance that it has been more than 1.7°F.

In spite of extensive uncertainties, both in the data and in the projections based on that data, researchers are increasingly confident about their ability to decipher the relationship between past activities and recent warming, to distinguish the effect of rising concentrations of greenhouse gases and changing land-use patterns from natural variability and other influences on climate, and to develop projections of the pace and ultimate magnitude and distribution of future warming and related changes.[8] The great majority of experts conclude that they cannot explain observed patterns of warming and related changes without considering emissions from human activities and that it is very likely that most of the warming is due to human activities.[9] Those experts also conclude that ongoing emissions at current or rising levels will continue to raise atmospheric concentrations and temperatures indefinitely.[10] Even immediate, dramatic reductions in emissions would not necessarily halt changes to which past emissions have already committed the climate: If concentrations were stabilized today by cutting emissions immediately to a small fraction of current levels, the average global temperature would gradually continue to rise—increasing by another 0.5°F to 1.6°F above recent levels by the end of this century, according to one study.[11]

The accumulation of gases in the atmosphere, changes in land use, and the resulting changes in climate are having further cascading effects. The oceans are absorbing CO_2 and heat from the atmosphere and will continue to do so for centuries, even if emissions are curtailed, changing the chemistry of the oceans and raising sea level.[12] Glaciers and ice caps are gradually melting in response to rising temperatures, also contributing to a rise in sea level.[13] Forests and soils are responding to elevated CO_2 concentrations and temperatures as well.[14] Moreover, the effects of greenhouse gases interact with and can be enhanced or offset in particular regions by the direct and indirect effects of emissions of different types of aerosols as well as by changes in land cover.[15]

The climate system's response to emissions and to changes in land use is thus a complex, gradually unfolding process involving both rapid changes and interactions among several slowly responding components that adjust at differing and uncertain rates. Once perturbed by a large increase in greenhouse-gas concentrations and extensive changes in land cover, the system will take centuries to respond fully, as the oceans warm, ice melts, and other parts of the system adjust.[16] The responses will not be uniform but will vary across regions and seasons. Furthermore, the response need not be entirely smooth or evenly paced but could involve rather abrupt shifts long after the processes have been triggered.[17]

Temperature trends can be masked to some extent by inherent short-term variability in global and regional temperatures—a phenomenon replicated by climate models. For example, between 1980 and 2007, the average annual global temperature grew at an average rate of about 0.03°F per year, but in one year (1997) it rose by 0.42°F, and in another (1999) it fell by even more.[18] Some recent projections suggest that annual temperatures may cool slightly over the next decade in Europe and

Box 1.
The Current State of the Art in Climate Modeling

Climate models have gradually but steadily improved in detail, in the range of phenomena that they include, and in their ability to replicate characteristics of the Earth's climate system.[1] Studies that measure and compare the ability of the current generation of models to simulate recent climate conditions—such as the extensive comparison of 23 of the most complex models in the *Fourth Assessment Report of the Intergovernmental Panel on Climate Change*—show that nearly all of them have improved in most respects. The most advanced models typically include linked representations of the atmosphere, oceans, sea ice, and land surface; most types of greenhouse gases and other relevant components of atmospheric chemistry; and characteristics of the carbon cycle.

The models replicate seasonal and large-scale regional variations in temperature and, to a lesser extent, precipitation; they also replicate large-scale ocean currents, large-scale ocean and climate oscillations, and storms and jet streams in the middle latitudes. However, for some phenomena, such as the dynamics of glaciers, the models remain in early stages of development. Moreover, because the global models have relatively coarse spatial resolutions, regional models of higher resolution are used to "downscale" the results of large models to analyze smaller-scale phenomena. The quality of such downscaling exercises is constrained by the limitations of and uncertainties in the global models.

The models plausibly replicate 20th-century climate trends when they are run with historical emissions of greenhouse gases, other types of emissions, and variations in natural forces, such as volcanic eruptions and fluctuations in solar energy. No model replicates those climate trends through variations in natural forces alone.

Because research groups vary in the way they represent uncertain aspects of the Earth's climate system, the models produce a range of results for many important climate indicators. Studies comparing the models find that the average of the models' simulations (referred to as the "ensemble-mean model") generally replicates features of the system better than does any single model. As a result, researchers have focused on ensemble-mean projections of climate change under various scenarios as being a type of "best guess" of likely changes, taking the range of model results as representing, to an extent, the uncertainty in researchers' current understanding of likely developments.[2]

However, that approach understates uncertainty because it overlooks the fact that each model incorporates its builders' best guesses for uncertain parameters (guesses that do not reflect the full range of uncertainty about them) and that no model includes all human influences on regional climates. To better analyze the full extent of uncertainty, researchers turn to simpler models that can be "tuned" to replicate the results of larger models but that can also be systematically varied to simulate other possible combinations of parameters.

---

1. This discussion draws primarily from David C. Bader and others, *Climate Models: An Assessment of Strengths and Limitations. Synthesis and Assessment Product 3.1. Report by the U.S. Climate Change Science Program and the Subcommittee on Global Change Research* (Washington, D.C.: U.S. Climate Change Science Program, 2008); and D. A. Randall and others, "Climate Models and Their Evaluation," in Susan Solomon and others, eds., *Climate Change 2007: The Physical Science Basis. Contribution of Working Group I to the Fourth Assessment Report of the Intergovernmental Panel on Climate Change* (Cambridge, U.K.: Cambridge University Press, 2007).

2. See Gerald A. Meehl and others, "Global Climate Projections," in Susan Solomon and others, eds., *Climate Change 2007: The Physical Science Basis. Contribution of Working Group I to the Fourth Assessment Report of the Intergovernmental Panel on Climate Change* (Cambridge, U.K.: Cambridge University Press, 2007), pp. 797–801.

North America, while the average global temperature remains fairly steady.[19] Aerosols from volcanic eruptions and other sources could temporarily result in regional or even global cooling. Such fluctuations are not inconsistent with an ongoing long-term warming trend brought about by human activities.

Longer-term scenarios involve considerable uncertainties that become more extensive the further into the future researchers attempt to project. Important uncertainties include the following:

■ *How population growth, technological developments, and economic change will influence land cover and the growth of emissions.* Projections of long-term changes in population, output, emissions, and land use are necessarily quite uncertain. For example, one analysis concludes that the global population will probably be around 8.4 billion at century's end (compared with nearly 6.8 billion today), but the study also projects a 10 percent chance that the population will be less than 5.6 billion and a 10 percent chance that it will be more than 12.1 billion.[20] Wide ranges of uncertainty also apply to projections of long-term changes in productivity, advances in energy technology, and the demand for fossil fuel and land use.[21] As a result, cumulative emissions of greenhouse gases and changes in land use over the course of the 21st century and beyond can be projected only as the extremely uncertain outcome of several very uncertain long-term trends.

■ *How rapidly the climate system will respond to accumulating greenhouse gases and other changes, and how much warming will ultimately occur (what is referred to as the climate's sensitivity).* The response to rising concentrations of greenhouse gases and other perturbations involves a range of feedbacks that can begin to develop immediately but, in some important cases, can take centuries or even millennia to fully unfold.[22] With respect to the short-term response, the most important uncertainties include the influence of clouds (which can enhance or offset warming, depending on where and when they form) and the effects of aerosols (which can not only offset warming directly but also influence cloud formation).[23] Longer-term uncertainties involve the uptake of CO_2 by the oceans and forests, the absorption of heat by the oceans, and the effect of warming on the continental ice sheets of Greenland and Antarctica.[24] Reasonable differences in how such processes are treated in climate models lead to large differences in the projected magnitude and distribution of warming and other changes, even when the models are all run with the same emissions scenario. Taking such uncertainties into account, the IPCC concluded that for any specific scenario, the change in average global temperature in 2100 was likely to be within a range of 40 percent below to 60 percent above the models' average.[25] That is, if all of the models were run with a specific emissions scenario, and the warming in 2100 (averaged across all models) was 5°F, then there are about two chances in three that the actual warming under such a scenario would be between 3°F and 8°F—and one chance in three that it could be higher or lower. There is substantial uncertainty about the amount of further warming over following centuries as well, because it will depend on how rapidly natural processes remove greenhouse gases from the atmosphere and how rapidly the system responds to the remaining gases.[26]

Uncertainties regarding the amount of future emissions and the climate system's response appear to make roughly comparable contributions to the overall uncertainty about the impact of warming over the 21st century.[27] Taken together, those uncertainties are sufficiently large that many experts have been reluctant to project a likely range of long-term changes in the global climate, and no firm consensus on such a range exists. Nevertheless, an analysis by one research group that has undertaken a particularly detailed treatment of both economic and scientific uncertainties provides a useful illustration of the range of possible outcomes over the 21st century. That group recently revised its projections upward substantially, concluding that if no actions are taken to reduce the growth of emissions, the median increase in the average annual global temperature over the course of the 21st century would be around 9°F, with a two-in-three chance that warming will be between 7°F and 11°F and a 9-in-10 chance that it will be between 6°F and 13°F.[28] By way of comparison, the transition from the depths of the last ice age (the Last Glacial Maximum, about 21,000 years ago) to the present climate involved a long-term increase in temperature of roughly 7°F to 13°F.[29] Even if global emissions were nearly eliminated over the next hundred years, one study indicates that temperatures would still rise by between 1°F and 5°F above recent levels.[30] (See Figure 1 for an illustration of historical and projected warming.)

Figure 1.
Historical and Projected Climate Change

(In degrees Celsius) (In degrees Fahrenheit)

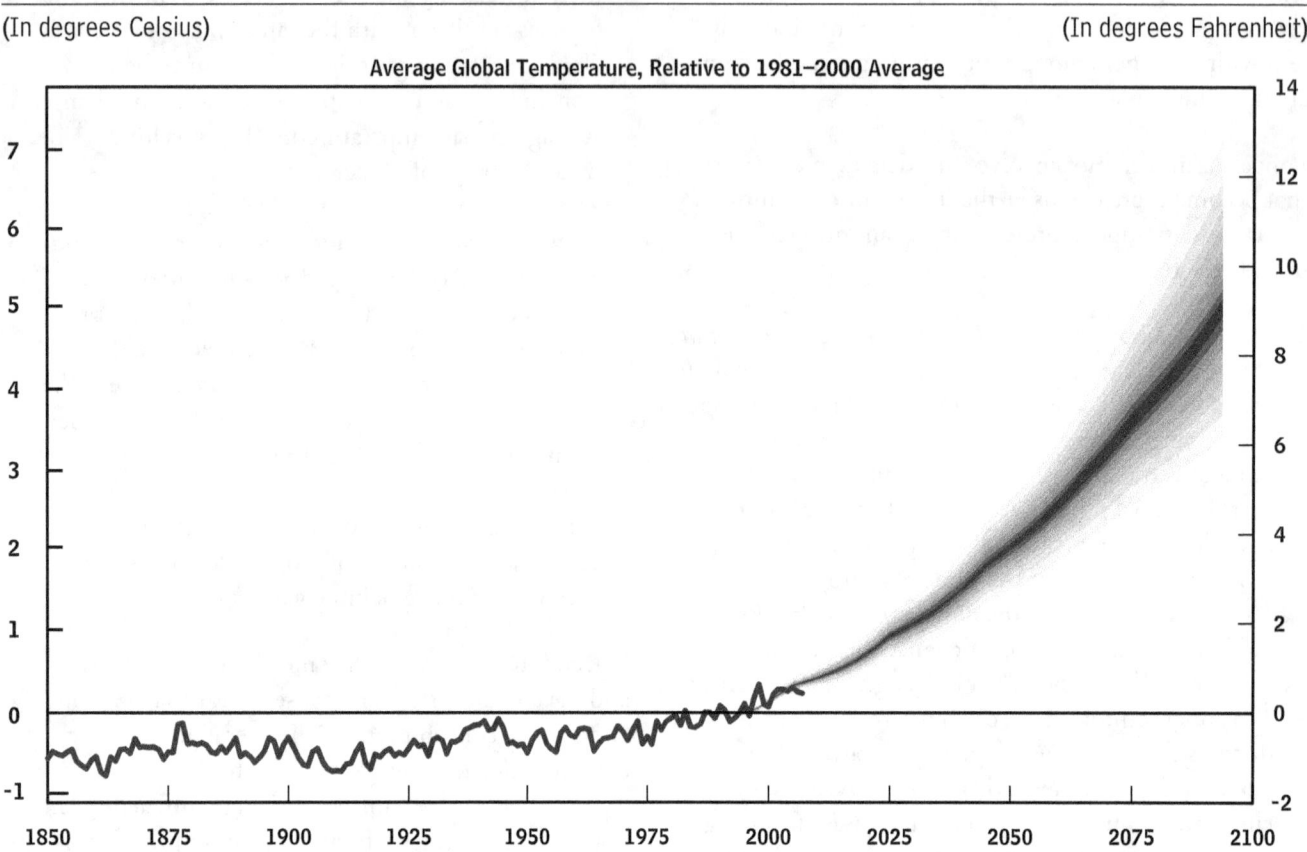

Source: Congressional Budget Office. Historical data are from the Hadley Centre for Climate Prediction and Research, http://hadobs.metoffice.com/hadcrut3/diagnostics/global/nh+sh/annual, and described primarily in P. Brohan and others, "Uncertainty Estimates in Regional and Global Observed Temperature Changes: A New Dataset from 1850," *Journal of Geophysical Research*, vol. 111 (June 24, 2006). The projection is based on data provided by Henry Jacoby, Massachusetts Institute of Technology, in a personal communication to CBO, December 22, 2008; the results are discussed in A.P. Sokolov and others, *Probabilistic Forecast for 21st Century Climate Based on Uncertainties in Emissions (Without Policy) and Climate Parameters,* Report No. 169 (Cambridge, Mass.: MIT Joint Program on the Science and Policy of Global Change, 2009), http://globalchange.mit.edu/files/document/MITJPSPGC_Rpt169.pdf.

Note: The projection, which is interpolated from decadal averages beginning in 1995, shows the possible distribution of changes in average global temperature as a result of human influence, relative to the 1981–2000 average and given current understanding of the climate. Under the Sokolov study's assumptions, the probability is 10 percent that the actual global temperature will fall in the darkest area and 90 percent that it will fall within the whole shaded area. However, actual temperatures could be affected by factors that were not addressed in the study (such as volcanic activity and the variability of solar radiation) and whose effects are not included in the figure.

Potential Impacts of Climate Change

Rising concentrations of greenhouse and other gases, changes in land use, and the resulting shifts in climate will result in many different kinds of impacts on widely differing scales and developing over widely varying periods of time. A number of such changes are already evident in the data, although, in many cases, the effects of human influences on the climate are very difficult to distinguish from other contributing causes.[31]

- A changing climate will involve changes in typical patterns of regional and seasonal temperature, rainfall, and snowfall, as well as changes in the frequency and severity of extreme events, such as heat waves, cold snaps, droughts, storms, and floods.[32] (See Figure 2 for an illustration of changes in averages and extremes of temperature and precipitation.) Regional climates in the United States are expected to become more variable, with more intense and more frequent extremes of

Figure 2.

Illustration of Changes in Averages and Extremes in Temperature and Precipitation

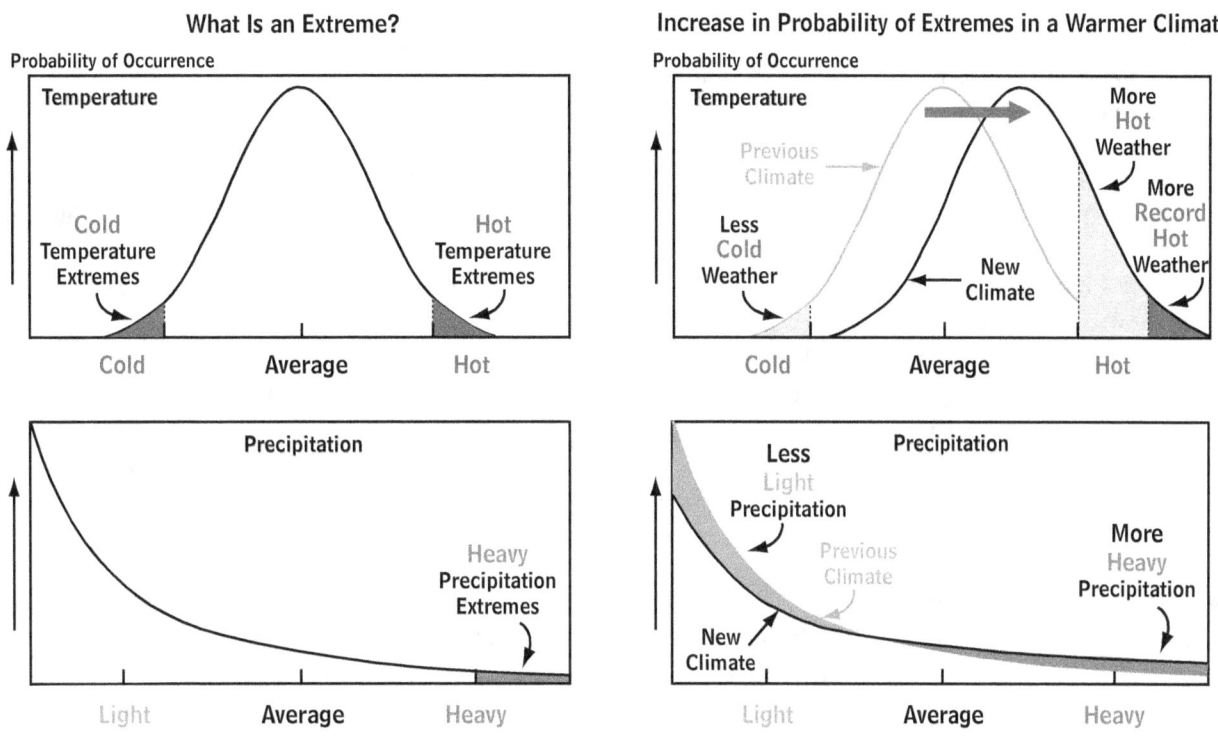

Source: Thomas R. Karl and others, eds., *Weather and Climate Extremes in a Changing Climate. Regions of Focus: North America, Hawaii, Caribbean, and U.S. Pacific Islands. A Report by the U.S. Climate Change Science Program and the Subcommittee on Global Change Research* (Washington, D.C.: Department of Commerce, National Oceanic and Atmospheric Administration, National Climate Data Center, June 2008), p. 2.

high temperature and rainfall. In general, extreme events tend to have disproportionately greater effects: A small percentage increase in hurricane wind speeds, for example, can greatly increase the potential damage.[33] Unfortunately, changes in the frequency and intensity of extreme events—especially precipitation—are also more difficult to simulate and project.[34]

- Some effects, including the melting of ice caps, a rise in sea level, and increasing acidity of the oceans, will unfold relatively gradually. Other effects could appear comparatively abruptly. Some extreme, abrupt changes—such as major shifts in ocean currents and regional patterns of rainfall—could occur unexpectedly, even centuries after emissions have been curtailed and concentrations have been stabilized.[35] A further complication is that changes, both gradual and extreme, will unfold at varying paces that, in some cases, are very difficult to project.

- In some cases, interactions among effects can amplify them in unpredictable ways. For example, rising sea level exacerbates coastal flooding from more intense storms.[36] Similarly, agricultural impacts are driven not only by changes in temperature and rainfall, but also by the enhancement of plant growth by higher atmospheric concentrations of CO_2 (the carbon fertilization effect), by the destructive effects of more ground-level ozone, and by more subtle effects, such as changes in susceptibility to pests and fire.[37]

- Yet another complicating factor is that the effects of climate change will occur along with demographic, economic, and environmental developments that are also difficult to project, that will have an impact on many of the same activities affected by climate change, and that, in many cases, will interact with (and mitigate or exacerbate) the effects of climate change in ways that are particularly difficult to untangle or project.[38]

Some changes, such as gradual shifts in average regional weather conditions, are likely to be relatively easy for a modern economy to anticipate and adapt to. Others, such as abrupt and unexpected changes in regional patterns of rainfall, could present much greater difficulties and impose much higher costs. Consideration of such complexities has gradually led researchers to shift from trying to gauge impacts precisely and to focus instead on characterizing vulnerabilities—the extent to which systems may be unable to cope with or adapt to adverse changes from multiple stresses.[39]

Effects on the Physical Environment

In general, projections of regional and local changes in climate are increasingly uncertain, the smaller the scale. Nevertheless, researchers find agreement among current climate models in consistently projecting certain types of changes in climate patterns over large geographic areas.[40] That agreement suggests that those projected changes are indeed likely to occur.

Temperature. Models suggest that as concentrations of greenhouse gases increase, warming will tend to be relatively greater at high latitudes in the Northern Hemisphere.[41] Warming will also tend to be greater over the continents than over the oceans and greater in the interiors of continents than near the coasts.[42] Average annual warming in the continental United States is likely to exceed the global average by roughly 25 percent to 35 percent—and by roughly 70 percent in Alaska.[43] Thus, if warming over the course of the 21st century is in the middle of the likely range indicated by the study discussed in the previous section (that is, about 9°F), average annual temperatures would rise by roughly 12°F in the continental United States and by nearly 16°F in Alaska. Further warming can be anticipated over the longer term even if emissions are curtailed by the end of this century.

In the central United States, the range of warming can be very roughly illustrated in terms of moving south—or, conversely, of having temperature patterns move north. At the low end of the study's likely range of 21st century warming (an increase of 7°F in average global temperatures), average annual temperatures would shift about two states north; that is, North Dakota's temperature range would become more like Nebraska's is today. At the high end of that likely range (that is, an increase of 11°F in average global temperatures), temperatures would shift about three states north over the 21st century: North Dakota would become almost as warm as Kansas is today, while central Kansas would become as warm as south Texas is today. (See Figure 3 for historical patterns of temperature and precipitation in the continental United States.)

Warming will also cause seasonal shifts throughout the world. It is very likely to result in shorter and milder winters, fewer and warmer cold days and nights, and fewer frost days; and it is also likely to yield longer and hotter summer conditions over most land areas, longer growing seasons, more and hotter warm days and nights, and more and hotter heat waves.[44] Northerly regions of the United States will tend to warm most in the winter, largely because declining snow and ice cover will increase the amount of heat absorbed by the land surface, while regions in the Southwest will tend to warm most in the summer.[45] Nevertheless, even northerly regions of the contiguous United States could warm substantially during the summer.[46]

Precipitation. Although projections of regional patterns of rainfall and snowfall are generally more uncertain than those for temperature, a warmer world is virtually certain also to experience more precipitation because warmer air can hold more moisture.[47] Indeed, precipitation has increased in most of the United States over the past half-century.[48] According to the results from current climate models, however, the increase in average precipitation over the course of the coming century will be unevenly distributed: Many regions and seasons that already have greater precipitation will tend to get more, while drier regions in the northern subtropics and midlatitudes will tend to get less.[49] Polar regions are likely to be an exception to that rule: They will tend to experience more precipitation despite being relatively arid at present.[50]

Projections of precipitation are particularly uncertain across a broad swath of northern Mexico and the continental United States. In those regions, a transition zone between relatively dry and wet climates moves north and south with the seasons, and the response of precipitation in that zone to a particular amount of warming is unclear.[51] Nevertheless, most models project that almost any amount of warming will cause much of the Southwest (as well as neighboring countries in much of Central America) to become more arid over the course of the 21st century—and at the high end of the range of possible warming, considerably more arid.[52]

Figure 3.
Historical Climate Patterns in the Continental United States, 1971 to 2000

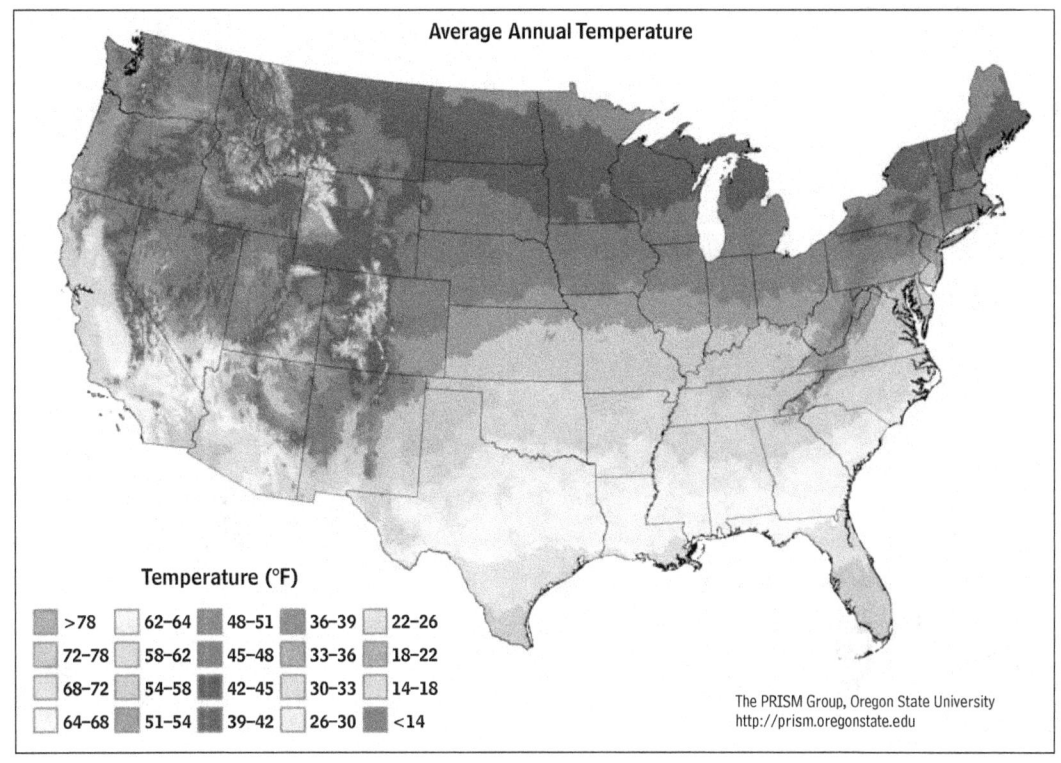

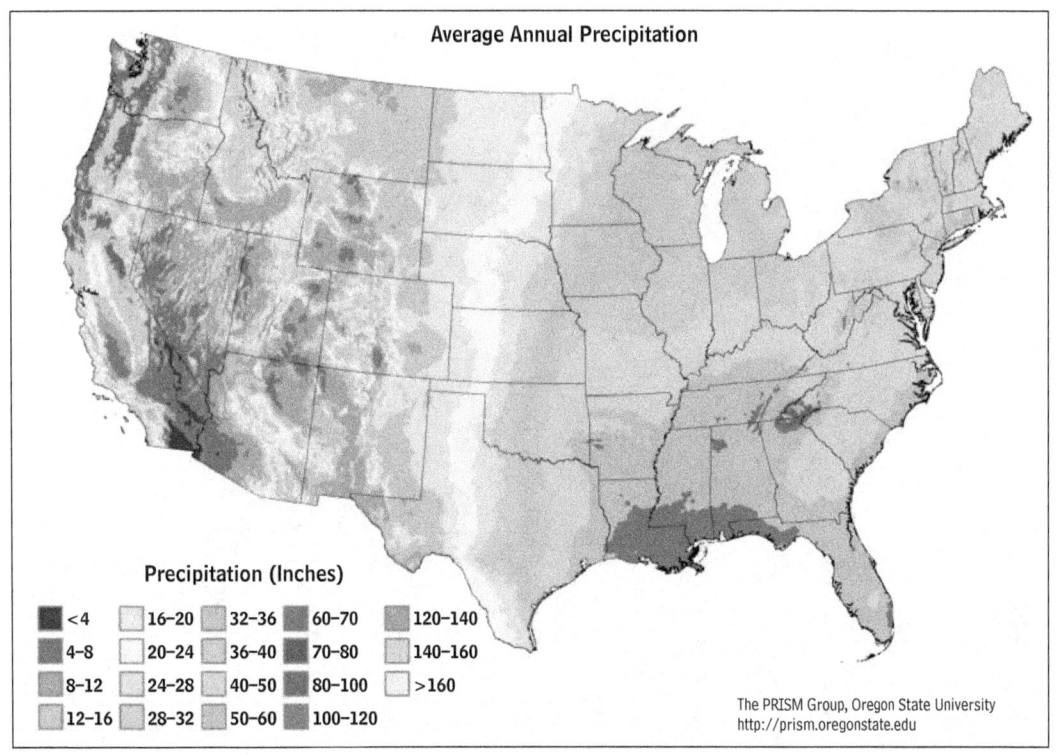

Continued

Figure 3. Continued
Historical Climate Patterns in the Continental United States, 1971 to 2000

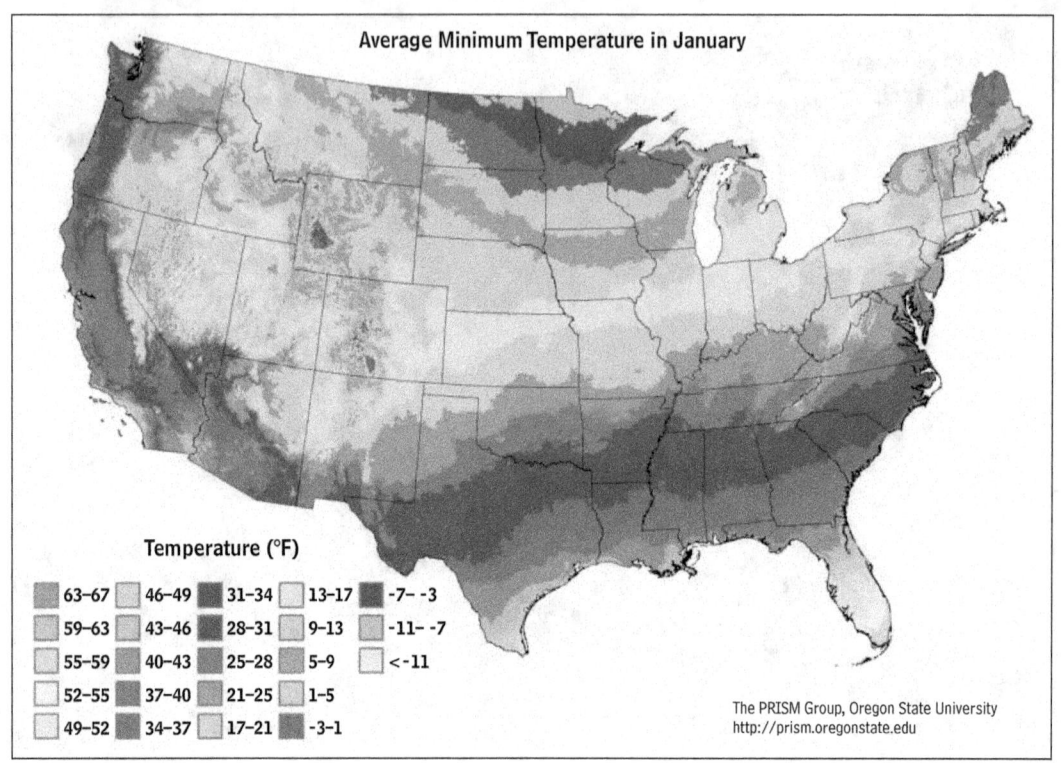

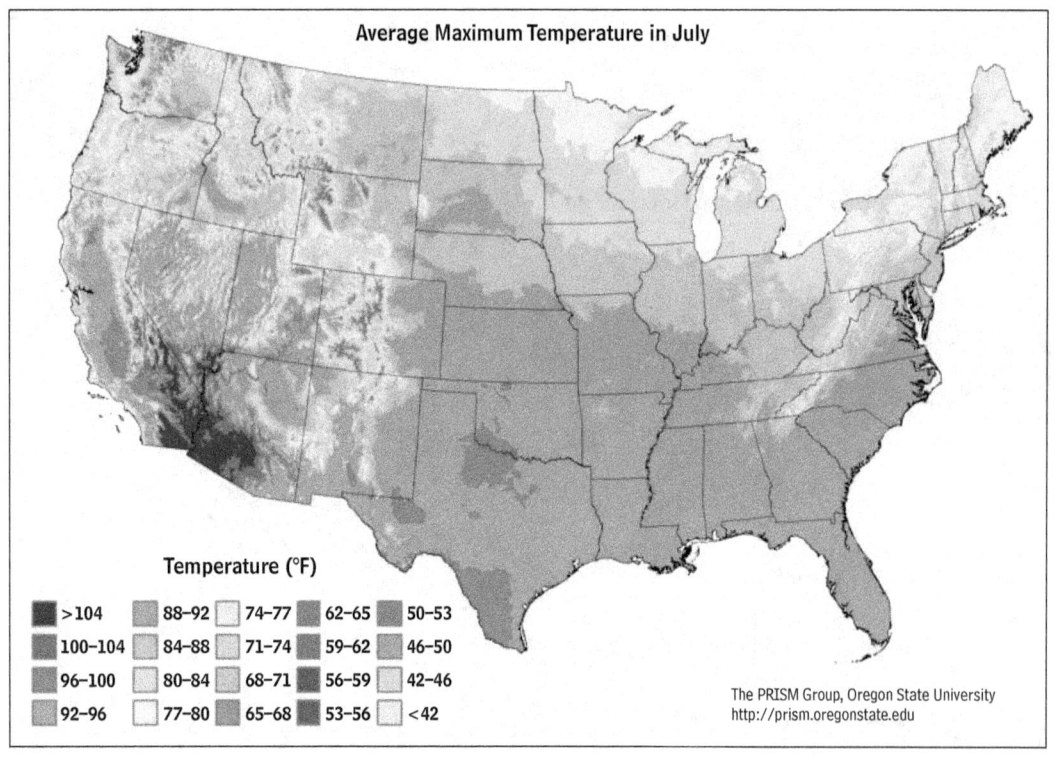

Source: The PRISM Group, Oregon State University; http://prism.oregonstate.edu.

Changes in seasonal weather patterns are more uncertain than annual averages, but as a rule, models show agreement in predicting that warmer winters will bring more precipitation in northerly regions of the United States, while summers will tend to become drier in southerly regions.[53] With warmer temperatures, less of the precipitation will fall as snow, and the snow will tend to melt earlier in the spring.[54] Warming even in the lower half of the projected range could dramatically shorten the snow season in much of the lower 48 states by the end of this century.[55] As discussed below, less snowfall and earlier melting could greatly reduce the supply of water during dry summer months in large parts of the western United States.

Patterns of rain and snowfall in specific locations also are likely to become more variable.[56] Severe summer thunderstorms are likely to become more frequent in large areas of the country.[57] In wet and dry regions alike, more intense and heavy rainfall will tend to be interspersed with longer relatively dry periods.[58] That pattern implies, somewhat paradoxically, that in a wetter, warmer world, many regions will experience greater risks both of flooding and of drought.

Regional patterns of precipitation also could be affected by changes in patterns of land use, by emissions and deposition of aerosols (and by their influence on cloud formation), and by the response of vegetation to changes in temperature and CO_2 concentrations.[59] Such regional responses are particularly difficult to project.

Cyclones, Typhoons, and Hurricanes. Many researchers have concluded that, on balance, warming over the 21st century will also tend to increase the overall size and intensity of tropical cyclones (called typhoons in much of the Pacific Ocean but called hurricanes in the Atlantic and eastern Pacific Oceans) by raising surface temperatures of the water in regions where such storms form.[60] A recent economic assessment concludes that the stronger hurricanes that would result from a temperature increase of 4.5°F—below the low end of the projected range of warming—would increase average annual U.S. hurricane damages by about $8 billion (in constant 2005 dollars), or about 0.06 percent of current gross domestic product.[61] However, future trends in the frequency and intensity of hurricanes remain very uncertain, especially in the North Atlantic, where nearly all hurricanes that reach the United States form.[62] An accurate assessment will require a better understanding of how large climate fluctuations (discussed below), which also influence the formation and persistence of hurricanes, are likely to change in a warming world.[63] Large storms outside of the tropics, such as nor'easters, may become less frequent but are likely to become more intense.[64]

Ocean Currents. Warming also could influence the Earth's major ocean currents, which are driven in part by wind patterns and which, in turn, strongly affect many regional patterns of climate. At present, warming appears to be strengthening the Antarctic Circumpolar Current, a strong westerly circulation around Antarctica in the Southern Ocean that is believed to account for the bulk of the wind-driven currents of the world's oceans.[65] In contrast, warming during the coming century is expected to slow the Atlantic meridional overturning circulation (MOC), a key component of the ocean circulation system that drives warm tropical waters far into the North Atlantic, into the deep ocean, and back to the south. That circulation is regulated in part by processes that increase the salinity and density of surface waters in the north Atlantic, causing them to sink into the deep ocean. Increasing rainfall and decreasing formation of sea ice could reduce surface salinity in the region, slowing the circulation process. No consensus currently exists regarding the likelihood that the MOC could shut down completely or the temperature increase that might trigger such an event.[66]

Climate Oscillations. Warming could be expected to have significant impacts on the large semiregular climate oscillations that dramatically affect weather patterns throughout the world, such as the El Niño/Southern Oscillation (ENSO), which results from interactions between ocean currents and wind patterns across the Pacific Ocean and causes significant variations in regional weather patterns from southern Africa to northeastern North America on a timescale of several years. Although researchers conclude that the overall global climate is likely to become more like that of a typical El Niño year, they currently cannot determine whether even dramatic warming would cause any significant change in the frequency or intensity of ENSO events.[67] Some evidence suggests that warming will tend to shift the typical effects of ENSO events in the United States—such as intense rainfall in the Southwest—northward and eastward over the 21st century.[68]

Sea Level. Warming will gradually raise sea level for many centuries after concentrations of greenhouse gases have been stabilized, partly by causing the oceans to expand,

and partly by melting glaciers, mountain ice caps, and parts of the Greenland and Antarctic ice sheets.[69] Warming of the oceans alone is projected eventually to raise sea level by roughly 4 to 13 inches per degree Fahrenheit of warming.[70] Moreover, even warming at the very low end of the projected range could result in melting of ice sheets that generates an unexpectedly rapid rise in sea level. The rate of that rise is very poorly understood, however, and constitutes one of the crucial scientific uncertainties surrounding climate change—so much so that the IPCC was unwilling to estimate a probable range for the contribution from melting ice sheets.[71]

Rising sea level will cause a gradual, progressive inundation and more rapid erosion of shorelines, coastal wetlands, and coastal infrastructure.[72] In the United States, most of the land vulnerable to inundation is in coastal Louisiana and east Texas, south Florida, the Pamlico-Albemarle Peninsula of North Carolina, and the Eastern Shore of Maryland.[73] Rising sea level will also tend to exacerbate the impacts of storm surges and floods in tidal zones: By one estimate, a three-foot rise in sea level would roughly double the incremental damages resulting from warming-induced increases in the intensity of hurricanes.[74] Because most large North American cities are on tidewater or rivers, they will face a gradual increase in the probability of what are currently unusually large storm surges. One study, for example, concludes that in the middle of the range of projected warming, New York City could experience what are now considered 100-year floods every three to four years by the end of the 21st century.[75]

Ocean Acidification. The world's oceans have partly offset the accumulation of CO_2 in the atmosphere by absorbing about one-third of historical emissions from human activities.[76] However, the absorption of CO_2 is also making ocean water more acidic. Ongoing emissions will continue to influence ocean chemistry: Even if the amount of emissions is at the low end of the projected range, seawater in some regions is expected to become more acidic during the 21st century than it has been in millions of years, with potentially serious effects on ocean ecosystems.[77]

Consequences for Biological Systems

By shifting climate zones in the United States about 50 miles poleward (and mountain climates about 300 feet higher) per degree Fahrenheit of warming, by shifting patterns of precipitation, and by changing ocean conditions, climate change will shift the types of vegetation and animals that thrive in specific locations.[78] Such environmental shifts are likely to affect unmanaged ecosystems, agriculture, commercial forestry, and fisheries.[79]

Ecosystems and Biodiversity. Ecosystems in the United States and around the world have already been measurably affected by rising concentrations of greenhouse gases, land-use changes, and changes in climate.[80] Under projected rates of warming over the next century, ongoing shifts in climate zones are very likely to further affect ecosystems. The impact on unmanaged ecosystems, including those in many of the world's regions of unusually rich biodiversity, could be particularly harmful.[81] In addition, shellfish, plankton, and corals face a highly uncertain threat from acidification of the world's oceans, which is making it more difficult for them to form and maintain their shells and skeletons.[82] In some regions, the pace of change in conditions on land and in the oceans could exceed species' capacity to respond by adapting or migrating.[83]

Warming is thus likely to place a substantial but very uncertain share of the Earth's species at risk of extinction—a share that is likely to grow with the rate and ultimate amount of emissions and warming.[84] Further uncertainties involve the combined effects on plant growth from shifts in climate, greater concentrations of CO_2, and higher levels of ozone; the pace at which various species will be able to disperse in the face of a shift in climate that, in geologic terms, is very rapid; and the simultaneous stresses caused by other widespread types of human disruption of natural habitats.[85] Given that about 2 percent of all known species of plants and animals have disappeared in the past few centuries and that the bulk of human population growth and resource use has occurred only since the mid-20th century, increasing pressures on biodiversity unrelated to climate are likely to interact with and exacerbate any stresses associated with climate change.[86]

Agriculture. Studies suggest that, in contrast to the difficulties for unmanaged ecosystems, the risks to agriculture in the United States are comparatively modest. Farmers in most regions of the United States can largely mitigate adverse impacts during the 21st century by adapting crops and techniques to changing climate conditions—for instance, by shifting northward the cultivation of specific crops and altering the timing of planting and harvesting in response to the earlier onset of warm weather and drier summer conditions.[87] If warming remains in the lower part of the projected range and if conventional air pollution controls mitigate the potential for damage from ozone, aggregate crop yields could even increase in the United States from the combined impact of changing regional climates and the carbon fertilization effect.[88]

Nevertheless, agriculture in some regions of the United States is likely to be adversely affected by climate change during the 21st century, mainly because of changes in rainfall rather than changes in temperature, and the effects are likely to be more adverse, the greater the amount of warming.[89] Although northerly regions of the country (as well as Canada) are likely to benefit from longer growing seasons and greater precipitation, more southerly regions tend to be much more sensitive to higher temperatures and would benefit less from increased precipitation.[90] Declining water supply could further affect agriculture in Western states. In other agricultural regions, greater and more intense precipitation would probably increase damage to crops from flooding.[91] Some effects of a changing climate—increasing variability of local climates as well as unpredictable abrupt climate shifts—could impair farmers' ability to determine what crops to cultivate and where and when to cultivate them. Over the coming century, the United States may become largely unsuitable for the cultivation of some crops, such as premium wine grapes, even as some areas of the country may become more suitable for others.[92]

Forestry. Studies generally conclude that warming and higher atmospheric concentrations of carbon dioxide will boost global forest growth and timber production over the course of the 21st century, although the effects on timber supply in specific regions will vary.[93] Warming and higher CO_2 concentrations have very likely already enhanced the growth of forests in regions of the United States where water is not a limiting factor.[94] Over the next century, changes in patterns of temperature and precipitation will shift the distribution of forest types throughout the country. The area of arid shrubland and steppe is likely to expand in the Southwest; in temperate regions, the ranges of forests that are mainly coniferous (dominated by evergreen trees such as spruce, fir, and pine) or deciduous (dominated by trees that shed their leaves each year) will expand; and the ranges of mixed forests in the northernmost temperate regions and in colder, more northerly boreal regions will shift northward and cover less land.[95] Even below the lower end of the projected range of warming, the tree line could shift northward by up to 250 miles in the Arctic over the course of the 21st century, dramatically increasing the Arctic forest area, shrinking the tundra, and releasing potentially large quantities of methane from melting tundra and from wetlands that would form on currently frozen ground.[96] Such shifts, in turn, could affect regional climates. Forests throughout the country are likely to be adversely affected by the northerly spread of insects and diseases, and, most importantly throughout much of the West, considerably greater risk of wildfires.[97] (Changes in CO_2 concentrations and climate have very likely already contributed to such developments, although their effects are very difficult to distinguish from those of other causes.)[98]

Fisheries. In rivers, reservoirs, lakes, estuaries, and oceans throughout the United States and its territories, rising temperatures will shift the geographic ranges of many fish northward, upriver, or deeper. Warm-water species will generally prosper, but even in the lower half of the likely range of warming, cold-water species such as salmon are likely to disappear from all but the deepest lakes in the continental United States over the course of the 21st century, while cod could disappear from Georges Bank off Cape Cod.[99]

Impacts on the Economy and Human Health

In addition to their effects on agriculture, forestry, and fisheries, changing climate conditions will affect a few industries and various types of infrastructure, such as water supply, and are also expected to affect human health in a variety of ways. Although those impacts could be significant on a regional scale, studies of the overall impact of climate change on measured economic activity in the United States have generally yielded estimates of economic costs that are modest in relation to overall economic activity, at least so long as warming is in the lower half of the projected range.

Water Supply. Changing precipitation patterns have important implications for water supply, especially in the arid Western states, where snowmelt supplies water through much of the dry summer season. During the 21st century, less winter snow accumulation and earlier spring melt will tend to decrease the amount of water available in summer months, significantly affecting the water supply in much of that region.[100] Under existing institutional arrangements, Western states will find it increasingly difficult to meet current levels of demand for water—not to mention the likely higher future levels needed by a larger population and a growing economy. The Colorado River flow is already declining, and recent research indicates that major water shortages are likely to occur in the Colorado River system within a matter of years.[101] Even in areas of the country that will experience a stable or increasing water supply overall, more intense rainstorms and other changes will tend to flush more materials into water supplies, while saltwater will tend to intrude into groundwater in coastal regions, increasing the cost of maintaining water quality.[102]

Infrastructure. While water supply appears to be the most important challenge, particularly in the West, various other types of infrastructure in the United States could be adversely affected by climate change in ways that would require significant expenditures for adaptation. The increased likelihood of flooding from a rising sea level, hurricanes, and intense precipitation may require investment in flood-control infrastructure along coasts and major rivers.[103] In Alaska, progressively thawing permafrost is expected to damage structures and roads.[104] However, warming is likely to reduce energy demand, all else being equal: In most American cities, warming is likely to raise the demand for energy for summer cooling but lower the demand for winter heating, which accounts for roughly twice as much energy use as cooling.[105]

Human Health. Warming is likely to affect human health in various ways. The most comprehensive national reviews of effects of climate change in the United States concluded that projected changes are likely to affect risks of morbidity (the incidence of disease) and mortality for several climate-sensitive aspects of human health, but that the overall net impact is uncertain.[106] For example, although heat waves are likely to become increasingly intense and potentially lethal, extreme cold snaps are likely to become less frequent. Recent research suggests that taken together, those changes in extremes are likely to yield only a very small increase in mortality over the course of the 21st century, especially once the potential for adaptation is taken into account.[107] Warmer climates could encourage the spread of a variety of insect-borne infectious diseases, however.[108] Warmer conditions could also be associated with higher levels of conventional air pollution in summer months but lower levels in winter, though the most recent national assessment concluded that projections of such changes were "somewhat speculative."[109] In general, most of the potential impacts on human health projected for the United States over the coming century appear to be modest, on net.

Aggregate Economic Impacts. Many of the projected impacts of climate change (such as damage from storms, hurricanes, and floods) will affect agriculture, forestry, and fishing; energy demand; and infrastructure. Because prices in those sectors are readily measured, evaluating the potential economic costs of specific effects of climate change is relatively straightforward. Thus, the significant uncertainty in estimates of economic costs arises more from uncertainty about the magnitude of future climate change and its physical impacts than from uncertainty about the economic costs of given physical impacts.

Despite the wide range of projected impacts of climate change over the course of the 21st century, published estimates of the economic costs of direct impacts in the United States tend to be modest.[110] Most of the economy involves activities that are not likely to be directly affected by changes in climate. Moreover, researchers generally expect the U.S. economy to grow dramatically over the coming century, mainly in sectors (such as information technology and medical care) that are relatively insulated from climate effects. Damages are therefore likely to be a smaller share of the future economy than they would be if they occurred today. As a consequence, a relatively pessimistic estimate for the loss in projected real (inflation-adjusted) U.S. gross domestic product is about 3 percent for warming of about 7°F by 2100.[111]

However, such estimates tend to mask larger losses in subsectors of the economy. Some sectors in certain regions are likely to bear sizable costs requiring significant adjustments and adaptations, and a few sectors in a few regions may be eliminated altogether. Even at the low end of the projected range of warming, for example, changing winter conditions would cut the Western ski season by up to four months and would virtually eliminate the Eastern snowmobiling season.[112]

In addition, most of the published studies do not include all of the potential costs of climate change to the country over the coming century and beyond. Most important, there are few detailed estimates of the costs of warming in the upper half of the projected range of 6°F to 13°F of warming during the 21st century.

Moreover, even for the levels of warming that have been examined, most of the estimates discussed above cover only a portion of the potential costs, for several reasons:

- *Nonmarket impacts.* Some types of impacts are very difficult to evaluate in monetary terms because they do not directly involve products that are traded in markets.[113] Although such difficulties apply to effects on human health and quality of life, they are particularly significant for biological impacts, such as loss of species' habitat, biodiversity, and ecosystem services (the various resources and processes that are supplied by natural ecosystems). Experts in such issues generally believe that those nonmarket impacts are much more likely to be negative than positive and could be large. It is particularly difficult to estimate how future generations, which are likely to be wealthier than current ones, will value such impacts relative to the much higher income that they are projected to have.

- *The potential for abrupt changes.* Experts believe that there is a small possibility that even relatively modest warming could trigger abrupt and unforeseen effects during the 21st century that could be associated with quite large economic costs in the United States. Two examples of such effects are shifts in ocean currents that could change weather patterns and affect agriculture over large areas, and rapid disintegration of ice sheets, which could dramatically raise sea level. The sources and nature of such abrupt changes, their likelihood, and their potential impacts remain very poorly understood.

- *Impacts outside the United States.* Most experts agree that populations in other countries—especially poor countries near the equator and bordering on desert zones—are likely to suffer the bulk of the damage from climate change during the 21st century.[114] Some countries could experience significant losses even at relatively low levels of warming—losses that would constitute a very small fraction of global economic output but that would nonetheless be catastrophic for the countries concerned. Some experts conclude that warming could result in "multiple chronic conditions" in some regions—for instance, simultaneous flooding from a rise in sea level, a reduced water supply, agricultural losses, and the spread of disease. Such conditions could seriously harm living standards that are already marginal in regions of Asia, Africa, and the Middle East, perhaps contributing to widespread political instability, with potentially large but very uncertain implications—including significant national security consequences—for the United States.[115]

The most comprehensive published study includes estimates of nonmarket damages as well as costs arising from the risk of catastrophic outcomes associated with about 11°F of warming by 2100. That study projects a loss equivalent to about 5 percent of U.S. output and, with substantially larger losses in a number of other countries, a loss of about 10 percent of global output.[116]

Policy Implications of Uncertainty

The uncertainties surrounding future emissions of greenhouse and other gases and land-use changes, the climate system's response to those developments, and the resulting impacts greatly complicate the crafting of a policy response. They make the climate outcome of any particular policy very difficult to determine and, conversely, make the appropriate policy to achieve any desired climate outcome very hard to predict.[117] Given current knowledge:

- *Any particular amount of emissions could result in much more or much less warming and associated impacts than expected.* Emissions in the middle of the range of projections discussed previously are most likely to result in 9°F of warming. However, that scenario has about a one-in-six chance of yielding less than 5°F of warming and about an equal chance of yielding more than 14°F. A policy that limited emissions with the goal of stabilizing atmospheric concentrations of greenhouse gases at roughly 650 parts per million of CO_2 equivalent (roughly 2.3 times the preindustrial CO_2 concentration of 280 parts per million) would significantly reduce warming, compared with a midrange scenario, but the policy could result in warming of anywhere between 3°F and nearly 6°F over the course of the 21st century (see Figure 4). Even if concentrations were stabilized at today's levels by dramatically reducing emissions, there is a small chance that a long-term temperature increase of nearly 4°F over the preindustrial level would nevertheless occur. Conversely, a fourfold increase in atmospheric CO_2 concentrations would have a small probability of yielding long-term warming of only about 6°F.

Figure 4.
Historical and Projected Climate Change, With and Without an Illustrative Policy Applied

(In degrees Celsius) (In degrees Fahrenheit)

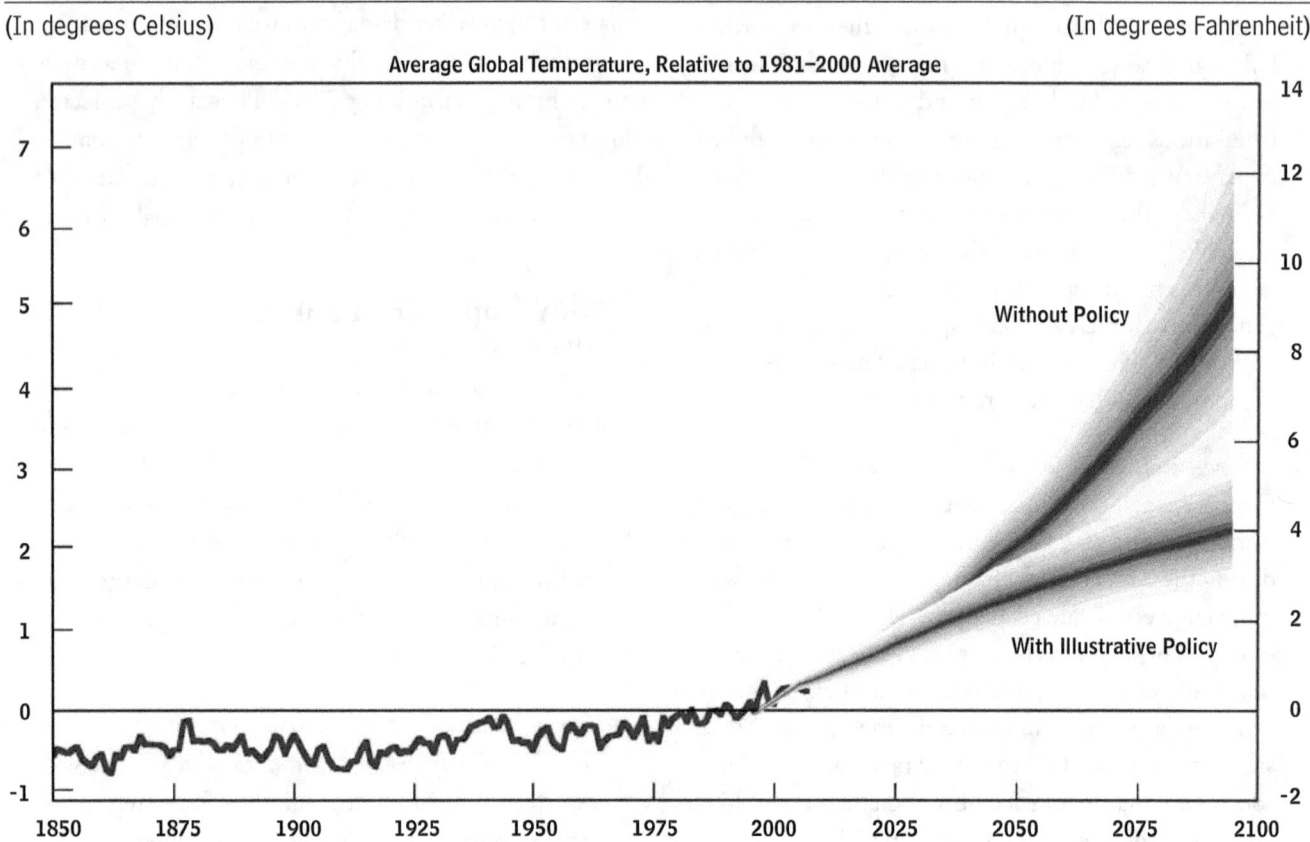

Source: Congressional Budget Office. Historical data are from the Hadley Centre for Climate Prediction and Research, http://hadobs.met office.com/hadcrut3/diagnostics/global/nh+sh/annual, and described primarily in P. Brohan and others, "Uncertainty Estimates in Regional and Global Observed Temperature Changes: A New Dataset from 1850," *Journal of Geophysical Research*, vol. 111 (June 24, 2006). The projections are based on data provided by Henry Jacoby, Massachusetts Institute of Technology, in a personal communication to CBO, December 22, 2008. The projection without policy is presented in A.P. Sokolov and others, *Probabilistic Forecast for 21st Century Climate Based on Uncertainties in Emissions (Without Policy) and Climate Parameters*, Report No. 169 (Cambridge, Mass.: MIT Joint Program on the Science and Policy of Global Change, 2009), http://globalchange.mit.edu/files/document/MITJPSPGC_Rpt169.pdf (in press in the *Journal of Climate*). The projection with an illustrative policy is the "Level 2" scenario discussed in L. Clarke and others, *Scenarios of Greenhouse Gas Emissions and Atmospheric Concentrations: Subreport 2.1A of Synthesis and Assessment Report 2.1 by the U.S. Climate Change Science Program and the Subcommittee on Global Change Research* (Washington, D.C.: Department of Energy, Office of Biological and Environmental Research, 2007).

Note: The projections, which are interpolated from decadal averages beginning in 1995, show the possible distribution of changes in average global temperature as a result of human influence, relative to the 1981–2000 average and given current understanding of the climate. The projection without policy assumes that no actions are taken to limit the growth of emissions. The projection with policy simulates the results over time of a cap on emissions that is designed to ensure that atmospheric concentrations of carbon dioxide remain below 550 parts per million. Under such a policy, cumulative global carbon dioxide emissions from human activities are assumed to be limited to 667 billion tons of carbon over the 21st century. The policy constrains emissions, not concentrations, so the resulting carbon dioxide concentrations over the 21st century grow to a median of 557 parts per million, with a 95 percent probability interval of 508 to 602 parts per million; concentrations of all greenhouse gases, summed using Global Warming Potential indices, have a 95 percent range of 599 to 714 parts per million of carbon dioxide equivalent. For each scenario, the probability is 10 percent that the actual global temperature will fall in the darkest area and 90 percent that it will fall within the whole shaded area. However, actual temperatures could be affected by factors that were not addressed in either scenario (such as volcanic activity and the variability of solar radiation) and whose effects are not included in the figure.

- *The amount of emissions that is compatible with a given climate goal is extremely uncertain.* For instance, suppose that policymakers wanted to limit total warming during the 21st century to no more than 5°F. If the climate's sensitivity to rising concentrations of greenhouse gases is at the low end of the likely range, policymakers could allow concentrations to more than double over their current level before stabilizing them. Little if any action would be required in the near term to meet that goal. However, if the climate is particularly sensitive, policymakers would have to sharply limit the growth of emissions over the next few decades to meet the same temperature goal.

Therefore, policies that target emissions or concentrations of greenhouse gases cannot guarantee specific climate outcomes; they can only shift the odds of those outcomes. Keeping concentrations to less than twice their preindustrial level—a commonly discussed target—would leave a more than 50 percent chance of ultimately exceeding a 5°F increase in average global temperature and a small chance even of exceeding 8°F.[118] Limiting concentrations further would reduce the chance of exceeding 5°F, but concentrations would have to be kept under about 420 parts per million of CO_2 equivalent to nearly eliminate that chance altogether. (With only the long-lived greenhouse gases taken into account, the current concentration is roughly 455 parts per million of CO_2 equivalent. However, with the offsetting effects of aerosols and other gases taken into account, the net current concentration is roughly 375 parts per million of CO_2 equivalent, with a very wide range of uncertainty around that estimate. See Figure 5 for approximate distributions of temperature under different scenarios for CO_2 concentrations.)

Given those uncertainties, crafting a policy response to climate change involves balancing two types of risks: the risks of limiting emissions to reach a temperature target and experiencing much more warming and much greater impacts than expected versus the risks of incurring costs to limit emissions when warming and its impacts would, in any event, have been less severe than anticipated. Climate policies thus have a strong element of risk management: Depending on the costs of doing so, society may find it economically sensible to invest in reducing the risk of the most severe possible impacts from climate change even if their likelihood is relatively remote.[119] In particular, the potential for unexpectedly severe and even catastrophic outcomes, even if unlikely, would justify more

Figure 5.

Uncertainty in the Climate's Response to Rising Atmospheric Concentrations of Carbon Dioxide

(Equilibrium change in average global temperature from the preindustrial level, in degrees Fahrenheit)

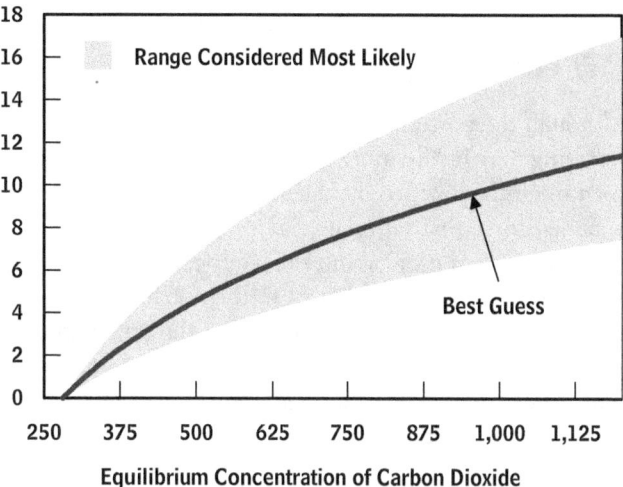

Source: Congressional Budget Office based on Table 10.8 in Gerald A. Meehl and others, "Global Climate Projections," in Susan Solomon and others, eds., *Climate Change 2007: The Physical Science Basis. Contribution of Working Group 1 to the Fourth Assessment Report of the Intergovernmental Panel on Climate Change* (Cambridge, U.K.: Cambridge University Press, 2007), p. 826.

Note: The figure shows the "best guess" and "most likely" range for the ultimate long-term amounts of warming that would result if, all else being equal, atmospheric concentrations of carbon dioxide were stabilized at specified levels. For example, if concentrations were stabilized at 750 parts per million by volume, the best guess is that the average global temperature would increase by nearly 8 degrees Fahrenheit above the preindustrial level, and there are about two chances in three (the "most likely" range) that the increase would be between 5 degrees and 11 degrees.

stringent policies than would result from simply balancing the costs of reducing emissions against the benefits of averting damages from the expected or most likely degree of warming.

Those insights have spurred some researchers who are particularly worried about low-probability but high-impact outcomes to call for limiting long-term warming to no more than 3°F to 5°F with a high degree of certainty. However, since about 1.4°F of warming has already occurred, and past emissions have made a substantial amount of further warming inevitable, limiting long-term warming to such levels with a substantial degree of certainty would probably require very dramatic and potentially very expensive curtailment of expected future emissions. There is a large difference in costs between a policy that leaves a 50 percent risk of warming exceeding 5°F and a policy that virtually eliminates that risk. In moving along the continuum of risk from the former to the latter, each increment of risk reduction is likely to come at an increasing price.

Although the possibility of very serious impacts from warming may justify more stringent near-term policies than might be justified by balancing expected costs and benefits alone, that rationale may shift over time as researchers learn more about the likely range of warming and related risks in the future. On the one hand, researchers may gradually learn that the climate's response to rising concentrations is much larger or the impacts much more severe than expected, requiring even more stringent policies than were initially adopted. On the other hand, researchers may discover that the climate's response or the impacts of its changes are relatively mild, in which case policies might be justifiably left as originally set.

The uncertainties in the link between emissions and climate change, coupled with the potential for improvements in future understanding of the climate problem, imply that policies that gradually raise the price of emissions with the expectation of meeting a cumulative target for emissions are likely to yield greater long-term net economic benefits than policies that impose increasingly stringent quantitative limits on emissions with the same cumulative target. Neither type of policy is likely to achieve a specific warming target, and policies of either type, if adopted, would be very likely to require adjustments and refinements as better information became available. Uncertainties may thus justify flexible mechanisms even though they may simultaneously justify relatively stringent policies.

Endnotes

1. See Intergovernmental Panel on Climate Change, *Climate Change 2007: Fourth Assessment Report of the Intergovernmental Panel on Climate Change,* vols. I, II, and III (Geneva, Switzerland: 2008); National Science and Technology Council, *Scientific Assessment of the Effects of Global Change on the United States* (May 2008); and the online library of the U.S. Climate Change Science Program, www.climatescience.gov/Library/default.htm.

2. Much of the research presented in the *Fourth Assessment Report* and other recent studies is concerned with documenting uncertainties in an increasingly standardized framework. For example, events that are considered to have a greater than 99 percent probability are referred to as "virtually certain"; "very likely" events are more than 90 percent probable; "likely" ones are greater than 66 percent probable; events that are "about as likely as not" have a probability of 33 percent to 66 percent; "unlikely" events have less than a 33 percent likelihood; "very unlikely" ones have less than a 10 percent probability; and "exceptionally unlikely" outcomes have less than 1 percent probability. This paper applies that framework where appropriate. See Susan Solomon and others, "Technical Summary," in Solomon and others, eds., *Climate Change 2007: The Physical Science Basis. Contribution of Working Group I to the Fourth Assessment Report of the Intergovernmental Panel on Climate Change* (Cambridge, U.K.: Cambridge University Press, 2007), pp. 22–23.

3. For more background information on climate change, see Congressional Budget Office, *The Economics of Climate Change: A Primer.* (April 2003).

4. See Solomon and others, "Technical Summary," p. 29.

5. Ibid., pp. 24–35.

6. Ibid., pp. 31–36.

7. Ibid., pp. 36–37. Temperatures and changes in temperature are presented here in degrees Fahrenheit (°F), but most scientific documents present them in degrees Celsius (°C). The formula for conversion of levels is °C = (0.56 * °F) - 32; the formula for changes in temperature is Δ°C = 0.56 * Δ°F.

8. Ibid., pp. 58–66.

9. Ibid., pp. 60–62.

10. Ibid., pp. 68–69.

11. Ibid., p. 70.

12. Ibid., pp. 26–27, 47–48, 77–79; and James Hansen and others, "Earth's Energy Imbalance: Confirmation and Implications," *Science,* vol. 308, no. 5727 (June 3, 2005), pp. 1431–1435.

13. See Solomon and others, "Technical Summary," pp. 42–46.

14. Ibid., pp. 26–27, 43–48.

15. See National Research Council, Climate Research Committee, Committee on Radiative Forcing Effects on Climate, *Radiative Forcing of Climate Change: Expanding the Concept and Addressing Uncertainties* (Washington, D.C.: National Academies Press, 2005); Daniel Rosenfeld and others, "Flood or Drought: How Do Aerosols Affect Precipitation?" *Science,* vol. 321, no. 5894 (June 2, 2008), pp. 1309–1313; and Gregg Marland and others, "The Climatic Impacts of Land Surface Change and Carbon Management, and the Implications of Climate-Change Mitigation Policy," *Climate Policy,* vol. 3 (2003), pp. 149–157.

16. Ibid., pp. 79–80.

17. See P.U. Clark and others, *Abrupt Climate Change, A report by the U.S. Climate Change Science Program and the Subcommittee on Global Change Research* (Reston, Va.: U.S. Geological Survey, 2008); R. B. Alley and others, "Abrupt Climate Change," *Science,* vol. 299, no. 5615 (March 28, 2003), pp. 2005–2010; and José Rial and others, "Nonlinearities, Feedbacks, and Critical Thresholds Within the Earth's Climate System," *Climatic Change,* vol. 65 (2004), pp. 11–38.

18. Calculated using data from the Met Office Hadley Centre for Climate Change, http://hadobs.metoffice.com/hadcrut3/diagnostics/global/nh+sh/annual.

19. See N. S. Keenlyside and others, "Advancing Decadal-Scale Climate Prediction in the North Atlantic Sector," *Nature,* vol. 453 (May 1, 2008), pp. 84–88.

20. Wolfgang Lutz, Warren C. Sanderson, and Sergei Scherbov, eds., *The End of World Population Growth in the 21st Century: New Challenges for Human Capital Formation and Sustainable Development* (London: Earthscan, 2004), p. 40. The estimate of current world population is from the U.S. Census Bureau's World Population Clock, www.census.gov/main/www/popclock.html.

21. For example, see Nebojša Nakićenović and Rob Swart, eds., *Emission Scenarios* (Cambridge, U.K.: Cambridge University Press, 2000).

22. See Solomon and others, "Technical Summary," p. 60; and K. L. Denman and others, "Couplings Between Changes in the Climate System and Biogeochemistry," in Susan Solomon and others, eds., *Climate Change 2007: The Physical Science Basis,* p. 534.

23. See Solomon and others, "Technical Summary," pp. 31–35; and D. A. Randall and others, "Climate Models and Their Evaluation," in Susan Solomon and others, eds., *Climate Change 2007: The Physical Science Basis,* p. 636.

24. See Gerald A. Meehl and others, "Global Climate Projections," in Susan Solomon and others, eds., *Climate Change 2007: The Physical Science Basis,* pp. 822–831.

25. Ibid., p. 810; and R. Knutti and others, "A Review of Uncertainties in Global Temperature Projections over the Twenty-First Century," *Journal of Climate,* vol. 21, no. 11 (June 1, 2008), pp. 2651–2663.

26. See Susan Solomon and others, "Irreversible Climate Change Due to Carbon Dioxide Emissions," *Proceedings of the National Academy of Sciences,* vol. 106, no. 6 (February 10, 2009), pp. 1704–1709.

27. For example, note that the ratio of highest to lowest best estimates of warming from different emissions scenarios presented by the IPCC is roughly comparable to the ratio of upper to lower bounds of the likely range of outcomes for any particular scenario. See Solomon and others, "Technical Summary," p. 70. Also see A. P. Sokolov and others, *Probabilistic Forecast for 21st Century Climate Based on Uncertainties in Emissions (Without Policy) and Climate Parameters,* Report No. 169 (MIT Joint Program on the Science and Policy of Global Change, 2009), p.26, http://globalchange.mit.edu/files/document/MITJPSPGC_Rpt169.pdf.

28. See Sokolov and others, *Probabilistic Forecast for 21st Century Climate.* These estimates represent a substantial upward revision from the Joint Program's previous estimates, which CBO cited in *The Economics of Climate Change: A Primer.* The revisions result mainly from a reappraisal of scientific uncertainties, particularly about the rate at which oceans will absorb heat from the atmosphere. The median level of projected warming is substantially greater than the warming expected to result from the IPCC's A1B emissions scenario, which is used extensively in the *Fourth Assessment Report* to illustrate the current state of scientific understanding of sectoral and regional impacts of warming. Another recent study that develops probabilistic projections of emissions reports a median warming of about 7°F during the 21st century, with a 9-in-10 probability that warming will be between 5°F and 9°F. However, that study does not consider emissions of any greenhouse gases except CO_2. See David von Below and Torsten Persson, *Uncertainty, Climate Change and the Global Economy,* Working Paper No. 14426 (Cambridge, Mass.: National Bureau of Economic Research, October 2008), www.nber.org/papers/w14426.

29. See E. Jansen and others, "Palaeoclimate," in Susan Solomon and others, eds., *Climate Change 2007: The Physical Science Basis,* pp. 435, 447.

30. See D. P. Van Vuuren and others, "Temperature Increase of 21st Century Mitigation Scenarios," *Proceedings of the National Academy of Sciences,* vol. 105, no. 40 (October 7, 2008), pp. 15258–15262.

31. See, for example, Solomon and others, "Technical Summary," pp. 35–54; and Peter Backlund, Anthony Janetos, and David Schimel, *The Effects of Climate Change on Agriculture, Land Resources, Water Resources, and Biodiversity in the United States. A Report by the U.S. Climate Change Science Program and the Subcommittee on Global Change Research* (Washington, D.C.: U.S. Environmental Protection Agency, 2008), p. 3.

32. For a discussion of the importance of the distinction between changes in average conditions and changes in extreme events, see Stanley A. Changnon, "Economic Impacts of Climate Conditions in the United States: Past, Present, and Future," *Climatic Change*, vol. 68 (2005), pp. 1–9.

33. See William D. Nordhaus, *The Economics of Hurricanes in the United States,* Working Paper No. 12813 (Cambridge, Mass.: National Bureau of Economic Research, December 2006), www.nber.org/papers/w12813.

34. See Randall and others, "Climate Models and Their Evaluation," pp. 627–629.

35. See Alley and others, "Abrupt Climate Change," p. 2008.

36. See C. B. Field and others, "North America," in M. L. Parry and others, eds., *Climate Change 2007: Impacts, Adaptation and Vulnerability. Contribution of Working Group II to the Fourth Assessment Report of the Intergovernmental Panel on Climate Change* (Cambridge, U.K.: Cambridge University Press, 2007), p. 630.

37. See Randall and others, "Climate Models and Their Evaluation," p. 631.

38. See T. R. Carter and others, "New Assessment Methods and the Characterisation of Future Conditions," in M. L. Parry and others, eds., *Climate Change 2007: Impacts, Adaptation and Vulnerability,* pp. 144–160; Field and others, "North America," p. 635; and Backlund, Janetos, and Schimel, *The Effects of Climate Change*, p. 3.

39. Hans-Martin Füssel and Richard J. T. Klein, "Climate Change Vulnerability Assessments: An Evolution of Conceptual Thinking," *Climatic Change*, vol. 75 (2006), pp. 301–329.

40. See Solomon and others, "Technical Summary," p. 71.

41. Ibid.

42. Ibid.

43. J. H. Christensen and others, "Regional Climate Projections," in Susan Solomon and others, eds., *Climate Change 2007: The Physical Science Basis*, pp. 852–856.

44. Ibid., p. 862; Thomas R. Karl and Gerald A. Meehl, "Executive Summary," in Thomas R. Karl and others, eds., *Weather and Climate Extremes in a Changing Climate. Regions of Focus: North America, Hawaii, Caribbean, and U.S. Pacific Islands. A Report by the U.S. Climate Change Science Program and the Subcommittee on Global Change Research* (Washington, D.C.: Department of Commerce, National Oceanic and Atmospheric Administration, National Climatic Data Center, 2008), pp. 4, 8; and Claudia Tebaldi and others, "Going to the Extremes: An Intercomparison of Model-Simulated Historical and Future Changes in Extreme Events," *Climatic Change*, vol. 79 (2007), p. 22. Such changes are already evident in most of North America and are considered likely to be linked to human influences on the climate.

45. Christensen and others, "Regional Climate Projections," p. 887.

46. Ibid., p. 890.

47. See Randall and others, "Climate Models and Their Evaluation," p. 591; and Meehl and others, "Global Climate Projections," pp. 750, 762, 768. Consistent with that projection, global precipitation over land has increased by about 2 percent over the past century. See Backlund, Janetos, and Schimel, *The Effects of Climate Change*, p. 14.

48. See Backlund, Janetos, and Schimel, *The Effects of Climate Change*, p. 149.

49. See Solomon and others, "Technical Summary," p. 75; and Christensen and others, "Regional Climate Projections," p. 890.

50. See Meehl and others, "Global Climate Projections," p. 750.

51. See Solomon and others, "Technical Summary," p. 75.

52. Ibid.; and Richard Seager and others, "Model Projections of an Imminent Transition to a More Arid Climate in Southwestern North America," *Science*, vol. 316, no. 5828 (May 25, 2007), pp. 1181–1184.

53. Christensen and others, "Regional Climate Projections," p. 891.

54. Ibid.

55. See Field and others, "North America," p. 634.

56. See Meehl and others, "Global Climate Projections," pp. 750–751.

57. Robert J. Trapp and others, "Changes in Severe Thunderstorm Environment Frequency During the 21st Century Caused by Anthropogenically Enhanced Global Radiative Forcing," *Proceedings of the National Academy of Sciences*, vol. 104, no. 50 (2007), pp. 19719–19723, www.pnas.org/cgi/doi/10.1073/pnas.0705494104.

58. See Meehl and others, "Global Climate Projections," pp. 750, 782; and Tebaldi and others, "Going to the Extremes," pp. 185–211.

59. See National Research Council, *Radiative Forcing of Climate Change*, pp. 5–6; Rosenfeld and others, "Flood or Drought," pp. 1309–1313; and R. A. Pielke Sr. and others, "An Overview of Regional Land-Use and Land-Cover Impacts on Rainfall," *Tellus*, vol. 59B (2007), pp. 587–601.

60. See Meehl and others, "Global Climate Projections," p. 751; and Karl and Meehl, "Executive Summary," p. 6.

61. See Nordhaus, *The Economics of Hurricanes in the United States*.

62. See Meehl and others, "Global Climate Projections," p. 788.

63. See Jeffrey P. Donnelly and Jonathan D. Woodruff, "Intense Hurricane Activity over the Past 5,000 Years Controlled by El Niño and the West African Monsoon," *Nature*, vol. 447, no. 7143 (May 24, 2007), pp. 465–468; and Gabriel A. Vecchi and Brian J. Soden, "Increased Tropical Atlantic Wind Shear in Model Projections of Global Warming," *Geophysical Research Letters*, vol. 34 (April 18, 2007).

64. See Meehl and others, "Global Climate Projections," pp. 788–789.

65. See J. R. Toggweiler and Joellen Russell, "Ocean Circulation in a Warming Climate," *Nature*, vol. 451 (January 17, 2008), pp. 286–288.

66. See Solomon and others, "Technical Summary," p. 72; Karl and Meehl, "Executive Summary," p. 6; and Kirsten Zickfeld and others, "Expert Judgments on the Response of the Atlantic Meridional Overturning Circulation to Climate Change," *Climatic Change*, vol. 82, no. 3-4 (June 2007), pp. 235–265.

67. See Solomon and others, "Technical Summary," p. 73; and Meehl and others, "Global Climate Projections," p. 779.

68. See Gerald A. Meehl and others, "Current and Future U.S. Weather Extremes and El Niño," *Geophysical Research Letters*, vol. 34 (October 19, 2007).

69. See Solomon and others, "Technical Summary," pp. 70–71, 80; and National Science and Technology Council, *Scientific Assessment of the Effects of Global Change on the United States*, p. 6.

70. See Susan Solomon and others, "Irreversible Climate Change Due to Carbon Dioxide Emissions," *Proceedings of the National Academy of Sciences*, vol. 106, no. 6 (February 10, 2009), pp. 1704–1709.

71. Solomon and others, "Technical Summary," p. 71.

72. M. L. Parry and others, "Technical Summary," in Parry and others, eds., *Climate Change 2007: Impacts, Adaptation and Vulnerability*, pp. 40–41, 44–45; and Field and others, "North America," p. 630.

73. See James G. Titus and Charlie Richman, "Maps of Lands Vulnerable to Sea Level Rise: Modeled Elevations Along the U.S. Atlantic and Gulf Coasts," *Climate Research*, vol. 18 (2001), pp. 205–228.

74. See National Science and Technology Council, *Scientific Assessment of the Effects of Global Change on the United States*, pp. 6–7; and Nordhaus, *The Economics of Hurricanes in the United States*.

75. See Field and others, "North America," p. 633.

76. Estimate based on an extrapolation from Christopher L. Sabine and others, "The Oceanic Sink for Anthropogenic CO_2," *Science*, vol. 305, no. 5682 (July 16, 2004), pp. 367–371.

77. See J. A. Raven and others, *Ocean Acidification Due to Increasing Atmospheric Carbon Dioxide* (London: Royal Society, 2005); and Andy Ridgwell and Richard E. Zeebe, "The Role of the Global Carbonate Cycle in the Regulation and Evolution of the Earth System," *Earth and Planetary Science Letters*, vol. 234, no. 3-4 (June 15, 2005), pp. 299–315.

78. See Wilfried Thuiller, "Climate Change and the Ecologist," *Nature*, vol. 448 (August 2, 2007), pp. 550–552.

79. See Parry and others, "Technical Summary," pp. 38–40, 44–45.

80. See Backlund, Janetos, and Schimel, *The Effects of Climate Change*, pp. 178–181.

81. See Jay R. Malcolm and others, "Global Warming and Extinctions of Endemic Species from Biodiversity Hotspots," *Conservation Biology*, vol. 20, no. 2 (April 2006), pp. 538–548. Thirty-four hotspots around the world are thought to contain more than 50 percent of all plant species and 42 percent of all terrestrial vertebrate species on 2.3 percent of the Earth's land area. The only hotspots in the United States are the California Floristic Province, comprising the California coast and central valley, and the northernmost Madrean Pine-Oak Woodlands in the Southwest. For further information on hotspots of biodiversity, see www.biodiversityhotspots.org/Pages/default.aspx.

82. See A. Fischlin and others, "Ecosystems, Their Properties, Goods, and Services," in Parry and others, eds., *Climate Change 2007: Impacts, Adaptation and Vulnerability*, p. 213; Raven and others, *Ocean Acidification Due to Increasing Atmospheric Carbon Dioxide*; Ridgwell and Zeebe, "The Role of the Global Carbonate Cycle in the Regulation and Evolution of the Earth System"; Victoria J. Fabry, "Marine Calcifiers in a High-CO_2 Ocean," *Science*, vol. 320, no. 5879 (May 23, 2008), pp. 1020–1022; and Kent E. Carpenter and others, "One-Third of Reef-Building Corals Face Elevated Extinction Risk from Climate Change and Local Impacts," *Science*, vol. 321, no. 5888 (July 23, 2008), pp. 560–563.

83. See John W. Williams, Stephen T. Jackson, and John E. Kutzbach, "Projected Distributions of Novel and Disappearing Climates by 2100 AD," *Proceedings of the National Academy of Sciences*, vol. 104, no. 14 (April 3, 2007), pp. 5738–5742.

84. See Fischlin and others, "Ecosystems, Their Properties, Goods, and Services," p. 213.

85. Parry and others, "Technical Summary," pp. 37–38; Inger Greve Alsos and others, "Frequent Long-Distance Plant Colonization in the Changing Arctic," *Science*, vol. 316, no. 5831 (June 15, 2007), pp. 1606–1609; and Ronald P. Neilson and others, "Forecasting Regional to Global Plant Migration in Response to Climate Change," *BioScience*, vol. 55, no. 9 (September 2005), pp. 749–759.

86. Documentation on extinction from the International Union for Conservation of Nature is available at www.iucnredlist.org/info/stats.

87. See, for example, John Reilly and others, "U.S. Agriculture and Climate Change: New Results," *Climatic Change*, vol. 57 (2003), pp. 43–69.

88. See John Reilly and others, *Global Economic Effects of Changes in Crops, Pasture, and Forests Due to Changing Climate, Carbon Dioxide, and Ozone*, Report No. 149 (MIT Joint Program on the Science and Policy of Global Change, 2007), http://globalchange.mit.edu/files/document/MITJPSPGC_Rpt149.pdf.

89. See Allison M. Thomson and others, "Climate Change Impacts for the Conterminous USA: An Integrated Assessment," *Climatic Change*, vol. 69, no. 1 (March 2005), "Part 3: Dryland Production of Grain and Forage Crops," pp. 43–65, and "Part 5: Irrigated Agriculture and National Grain Crop Production," pp. 89–105.

90. See Raymond P. Motha and Wolfgang Baier, "Impacts of Present and Future Climate Change and Climate Variability on Agriculture in the Temperate Regions: North America," *Climatic Change*, vol. 70, no. 1–2 (May 2005), pp. 137–164; and Robert Mendelsohn and Michelle Reinsborough, "A Ricardian Analysis of U.S. and Canadian Farmland," *Climatic Change*, vol. 81. no. 1 (March 2007), pp. 9–17.

91. Cynthia Rosenzweig and others, "Increased Crop Damage in the US from Excess Precipitation Under Climate Change," *Global Environmental Change*, vol. 12, no. 3 (October 2002), pp. 197–202.

92. See Katharine Hayhoe and others, "Regional Assessment of Climate Impacts on California Under Alternative Emission Scenarios—Key Findings and Implications for Stabilisation," Chapter 24 in *Avoiding Dangerous Climate Change* (Cambridge, U.K.: Cambridge University Press, 2006); and M. A. White and others, "Extreme Heat Reduces and Shifts United States Premium Wine Production in the 21st Century," *Proceedings of the National Academy of Sciences*, vol. 103, no. 30 (July 25, 2006), pp. 11217–11222.

93. See M. G. Ryan and S. R. Archer, "Land Resources: Forests and Arid Lands," in Backlund, Janetos, and Schimel, *The Effects of Climate Change*, pp. 79–80; and W. E. Easterling and others, "Food, Fibre, and Forest Products," in Parry and others, eds., *Climate Change 2007: Impacts, Adaptation and Vulnerability*, pp. 288–290.

94. See Backlund, Janetos, and Schimel, *The Effects of Climate Change*, p. 119.

95. See R. César Izaurralde and others, "Climate Change Impacts for the Conterminous U.S.A.: An Integrated Assessment—Part 6: Distribution and Productivity of Unmanaged Ecosystems," *Climatic Change*, vol. 69, no. 1 (March 2005), pp. 107–126.

96. See Fischlin and others, "Ecosystems, Their Properties, Goods, and Services," pp. 230–232; Jed O. Kaplan and Mark New, "Arctic Climate Change with a 2°C Global Warming: Timing, Climate Patterns, and Vegetation Change," *Climatic Change*, vol. 79, no. 3–4 (December 2006), pp. 213–241; and Clark and others, *Abrupt Climate Change*, p. 16.

97. See Field and others, "North America," p. 631; and Backlund, Janetos, and Schimel, *The Effects of Climate Change*, pp. 7–8.

98. See Ryan and Archer, "Land Resources," pp. 78, 118.

99. See Field and others, "North America," pp. 631–632; Michael Fogarty and others, "Potential Climate Change Impacts on Atlantic Cod (Gadus Morhua) off the Northeastern U.S.A.," *Mitigation and Adaptation Strategies for Global Change*, vol. 13, no. 5–6 (June 2008), pp. 453–466; and Heinz G. Stefan, Xing Fang, and John G. Eaton, "Simulated Fish Habitat Changes in North American Lakes in Response to Projected Climate Warming," *Transactions of the American Fisheries Society*, vol. 130, no. 3 (May 2001), pp. 459–477.

100. See Field and others, "North America," p. 627; and National Science and Technology Council, *Scientific Assessment of the Effects of Global Change on the United States* (May 2008), pp. 12–13.

101. See Tim P. Barnett and David W. Pierce, "When Will Lake Mead Go Dry?" *Water Resources Research*, vol. 44, no. 3 (March 2008).

102. See Field and others, "North America," p. 629; and National Science and Technology Council, *Scientific Assessment of the Effects of Global Change on the United States*.

103. Field and others, "North America," pp. 632–633.

104. Ibid., p. 632; and Larry D. Hinzman and others, "Evidence and Implications of Recent Climate Change in Northern Alaska and Other Arctic Regions," *Climatic Change*, vol. 72, no. 3 (October 2005), pp. 251–298.

105. See National Science and Technology Council, *Scientific Assessment of the Effects of Global Change on the United States*, pp. 16–17; Olivier Deschênes and Michael Greenstone, *Climate Change, Mortality, and Adaptation: Evidence from Annual Fluctuations in Weather in the US*, Working Paper No. 13178 (National Bureau of Economic Research, June 2007), www.nber.org/papers/w13178; and Department of Energy, Energy Information Administration, *Annual Energy Outlook 2008*, DOE/EIA-0383 (2008), pp. 123, 125, www.eia.doe.gov/oiaf/aeo/pdf/0383(2008).pdf.

106. See J. L. Gamble and others, "Executive Summary," in Gamble and others, *Analyses of the Effects of Global Change on Human Health and Welfare and Human Systems. A Report by the U.S. Climate Change Science Program and the Subcommittee on Global Change Research* (Washington, D.C.: U.S. Environmental Protection Agency, 2008), pp. ES 5-7; National Science and Technology Council, *Scientific Assessment of the Effects of Global Change on the United States*, pp. 14–16; National Assessment Synthesis Team, *Climate Change Impacts on the United States: The Potential Consequences of Climate Variability and Change* (Cambridge, U.K.: Cambridge University Press, 2001), pp. 102–107; and Kristie L. Ebi and others, "Climate Change and Human Health Impacts in the United States: An Update on the Results of the U.S. National Assessment," *Environmental Health Perspectives*, vol. 114, no. 9 (2006), pp. 1318–1324.

107. See Mercedes Medina-Ramón and Joel Schwartz, "Temperature, Temperature Extremes, and Mortality: A Study of Acclimatisation and Effect Modification in 50 United States Cities," *Occupational and Environmental Medicine*, vol. 64 (2007), pp. 827–833; and Deschênes and Greenstone, *Climate Change, Mortality, and Adaptation*.

108. See Field and others, "North America," p. 632.

109. Ibid.; and National Science and Technology Council, *Scientific Assessment of the Effects of Global Change on the United States*, p. 15.

110. See Dale W. Jorgenson and others, *U.S. Market Consequences of Global Climate Change* (Arlington, Va.: Pew Center on Global Climate Change, 2004); Robert Mendelsohn and James E. Neumann, "Synthesis and Conclusions," in Robert Mendelsohn and James E. Neumann, eds., *The Impact of Climate Change on the United States Economy* (Cambridge, U.K.: Cambridge University Press, 2004), pp. 315–331; William D. Nordhaus and Joseph Boyer, *Warming the World: Economic Models of Global Warming* (Cambridge, Mass.: MIT Press, 2000); and Richard S. J. Tol, "Estimates of the Damage Costs of Climate Change—Part II: Dynamic Estimates," *Environmental and Resource Economics*, vol. 21, no. 2 (February 2002), pp. 135–160.

111. See Jorgenson and others, p. 36.

112. See Field and others, "North America," p. 634.

113. See Backlund, Janetos, and Schimel, *The Effects of Climate Change*, pp. 3–4.

114. See Parry and others, "Technical Summary," pp. 59–63.

115. See, for example, Kurt M. Campbell and others, *The Age of Consequences: The Foreign Policy and National Security Implications of Global Climate Change* (joint publication of the Center for Strategic and International Studies and the Center for a New American Security, 2007), www.csis.org/media/csis/pubs/071105_ageofconsequences.pdf; and CNA Corp., *National Security and the Threat of Climate Change* (Alexandria, Va., 2007), http://securityandclimate.cna.org/report/SecurityandClimate_Final.pdf.

116. See Nordhaus and Boyer, *Warming the World*, pp. 95–96.

117. For more background information on the policy implications of uncertainty, see Congressional Budget Office, *Uncertainty in Analyzing Climate Change: Policy Implications* (January 2005).

118. Different greenhouse gases persist in the atmosphere for different periods of time and have different warming characteristics. To simplify matters, researchers commonly refer to emissions and atmospheric concentrations of different greenhouse gases in terms of the amount of CO_2 that would cause an equivalent amount of warming over an arbitrary period of time—typically 100 years. Researchers often discuss concentration goals with reference to the preindustrial level of CO_2 (about 280 parts per million, or ppm). For example, a commonly mentioned goal is to limit the increase in concentrations to 550 ppm of CO_2 equivalent (CO_2e), or about twice the preindustrial level of CO_2, even though there were other greenhouse gases in the atmosphere before industrialization as well. The doubling limit would most likely constrain the long-term increase in average global temperature to slightly more than 5°F, but the likely outcome could be anywhere between 3°F and 8°F.

119. The risk management aspect of climate policy is more analogous to buying air bags than buying insurance, for two reasons. One is that climate policy involves making investments to reduce the risk of adverse outcomes rather than to ensure compensation in the event of such outcomes. The other reason is that unlike the case of automobile insurance, in which millions of drivers can pool their risks to compensate those who actually have accidents, climate change involves only one outcome on one planet. In that respect, no risk pooling is feasible. Thus, for society as a whole, expenditures to avert the greatest risks associated with warming could not be used to compensate anyone if those risks ultimately proved to be relatively mild.